AF207026

# THE POTENTIAL OF ARTIFICIAL INTELLIGENCE

by Jennifer Kaul

BrightPoint Press

San Diego, CA

© 2025 BrightPoint Press
an imprint of ReferencePoint Press, Inc.
Printed in the United States

For more information, contact:
BrightPoint Press
PO Box 27779
San Diego, CA 92198
www.BrightPointPress.com



LIBRARY OF CONGRESS CATALOGING-IN-PUBLICATION DATA

Name: Kaul, Jennifer, author.
Title: The Potential of Artificial Intelligence / by Jennifer Kaul.
Description: San Diego, CA: ReferencePoint Press, Inc., 2025 | Series: Focus on Artificial Intelligence | Audience: Grade 6 to 12 | Includes bibliographical references and index.
Identifiers: ISBN: 9781678209520 (hardcover) | ISBN: 9781678209537 (eBook)
The complete Library of Congress record is available at www.loc.gov.

# CONTENTS

# AT A GLANCE

- Artificial intelligence (AI) is the ability of a computer to do tasks usually done by humans.

- With pattern recognition, AI can collect, organize, and analyze large amounts of data.

- AI has the potential to positively impact daily life, safety, health, and the environment.

- Smart homes collect information on homeowners to tailor their home environment to their wants and needs, saving them time and money.

- AI could protect people by scanning for weapons in schools, concert venues, and public places. This could help prevent crimes.

- In the medical field, AI could be used to diagnose and treat diseases.

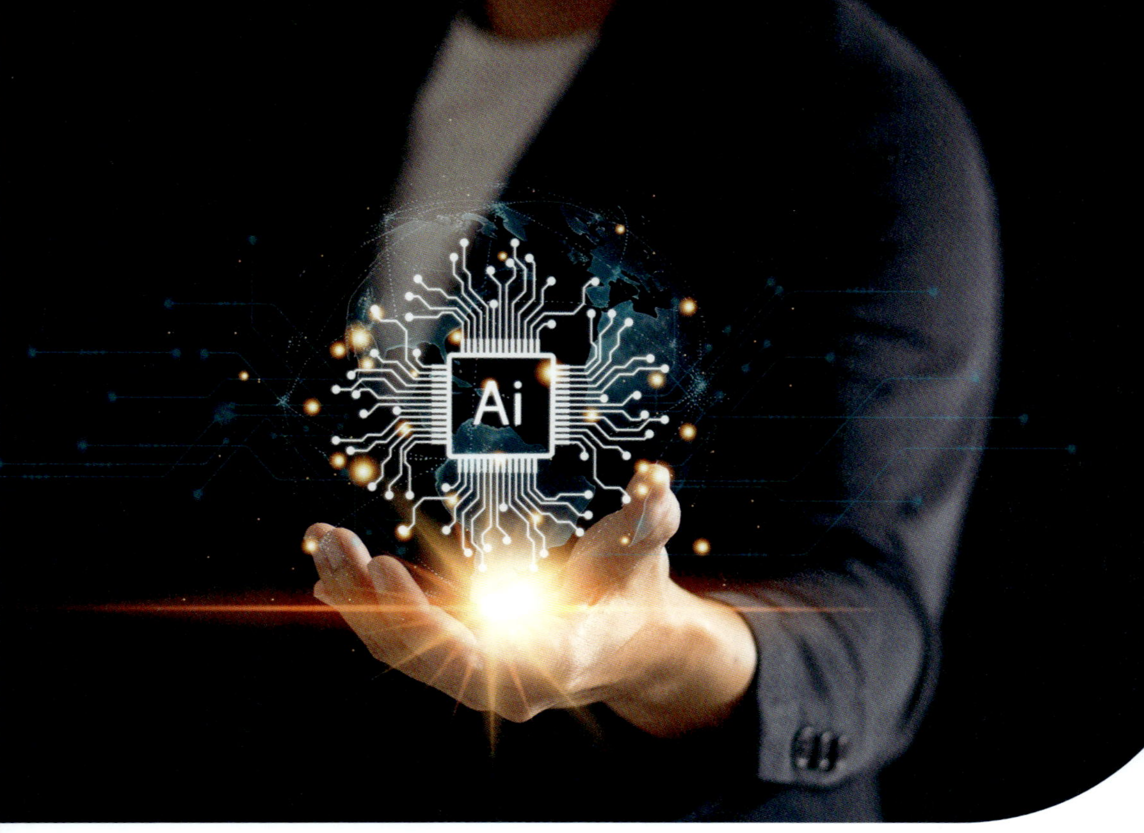


- On social media platforms and other online spaces, AI could detect and remove hate speech and fake news.

- Through improved recycling and agricultural practices, AI can reduce waste.

- AI can help protect land, water, and animals by using cameras, satellites, and sensors to track weather trends and watch for poachers.

# THE FUTURE'S POTENTIAL

It is a Friday afternoon in the year 2042. When the school bell rings, Asha is ready for the weekend. She hops on a self-driving bus to go home. Cameras on the bus scan her face when she boards. Other cameras work with **sensors** while the bus is driving. These sensors see people and cars and help the bus avoid them. They keep Asha and the other riders safe.

Facial recognition software may be more commonly used in the future to grant people access to buildings and services such as public transit.

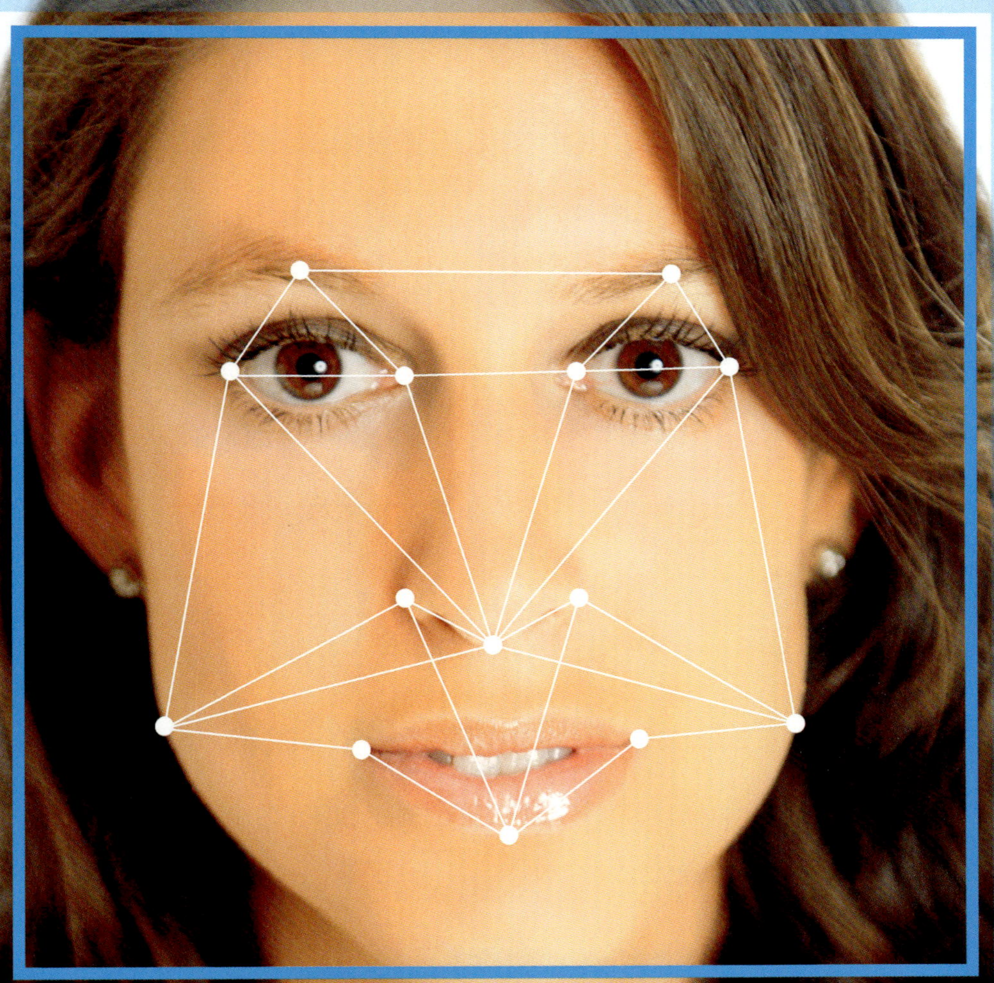

00000 000 0 00000000 00100111
0000 0 0000 0 0 0 00  100010111

010011000010 01000111000110
00101110101000011110101010
1101000010 10 11111000001001

IDENTITY PROTECTION

Name:

Password:

00000 000 0 00000000 00100111
0000 0 0000 0 0 0 00  100010111

010011000010 01000111000110
00101110101000011110101010

Asha arrives home. She is ready to relax.
But she has a basketball game the next
day. Her jersey needs to be washed. Still,
she plops down on the couch. Asha uses
her phone to ask her family's robot to sort
the laundry. She will start the wash later.
Then she turns on the TV. It recognizes
Asha. It streams the next episode of her
favorite show.

After dinner with her family, Asha gets
ready for bed. She brushes her teeth. She
looks in the mirror. A camera scans her
skin. The mirror has software that looks for
signs of illness. Her toilet uses sensors, too.
It scans her waste. She is low on some key
vitamins and minerals. An alert is sent to
the smart home's control center. It will order
some vitamins.

Asha's bed senses when she lies down. It sends an alert to her smart home network. This cools her room and dims her lights. It even streams calming music from Asha's personal playlist. Asha goes to sleep. Her smart home charges using energy saved from its solar panels.

**It's predicted that in the future, homeowners will no longer need an app to adjust the lighting, temperature, and music in a room. With smart homes this will happen automatically.**

**AI has the potential to help doctors.**

This story is fictional. But it is based on predictions about the future. Experts think artificial intelligence (AI) will advance in many of these ways in the future.

## WHAT IS AI?

AI is the ability of a computer to do tasks usually done by humans. AI starts by gathering data. It looks for patterns in the data. Then it applies what it learns to new data. This is called **machine learning (ML)**.

AI has the potential to improve life in many ways. It can save people time and keep them safe. It can keep people healthy. AI can protect the environment as well. These are just some of the many ways AI could change life for the better.

# CHAPTER ONE

# MAKING LIFE EASIER

AI could make people's lives easier. It is already used to complete basic tasks. For example, AI pulls up search results online. It creates playlists. It suggests products for people to purchase. This is just the beginning.

AI is already helping with more complex tasks. It will keep building on these skills. AI gathers and sorts through data. It can make decisions for people based on this data.

**AI can make unique playlists featuring music styles that match songs the listener has favored in the past.**

AI could even help people connect with new friends.

## AI'S EFFICIENCY

AI has been trained on data found on the internet. This includes writing and research by experts in many different fields, including medicine. With this data, AI can create almost anything. It can write reports and make artworks. It can provide

## What Is ChatGPT?

ChatGPT is one example of AI. It is a chatbot created by the company OpenAI. It was launched in November 2022. ChatGPT can answer questions in a way that makes it sound human. It can also perform tasks. It can write a poem, a report, or a résumé.

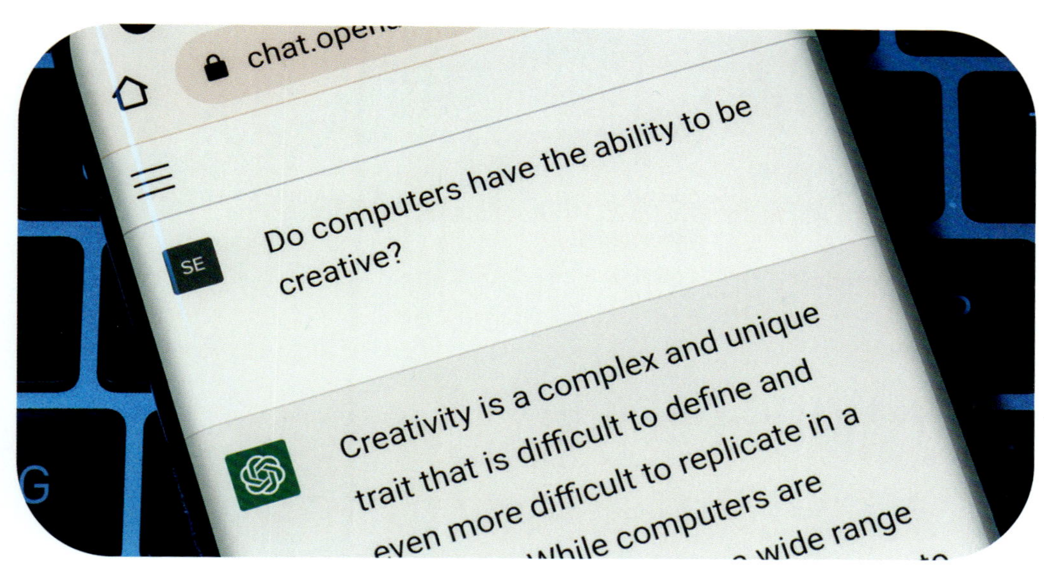



**Chatbots can write responses to questions and craft text.**

customer service. AI can solve math problems. It can do many of these tasks faster than people.

Human workers need breaks. But AI can work 24 hours a day. It does not get tired or hungry. It can perform at the same level all the time. This means AI gets more work done. This saves time.

AI can do jobs that many think are repetitive or boring. Examples might be entering data or sending emails. This could give people time to focus on more inspiring projects.

## MAKING DECISIONS

AI is very good at researching. It can gather and sort through a lot of information. AI can help people learn and make decisions. An applicant tracking system (ATS) is a form of AI. Some companies use ATS to scan job résumés. It rejects people who are not good fits for a job. This happens to up to 75 percent of applicants. In 2018, 67 percent of hiring managers felt ATS made their job easier.

AI can gather a lot of data in a short time. Then AI can quickly organize and understand it. It does this using pattern recognition. AI uses the patterns it finds

# AMERICAN OPINION ON USING AI TO MAKE HIRING DECISIONS

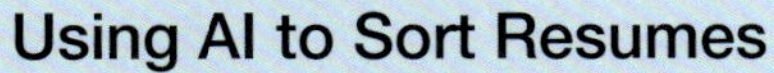

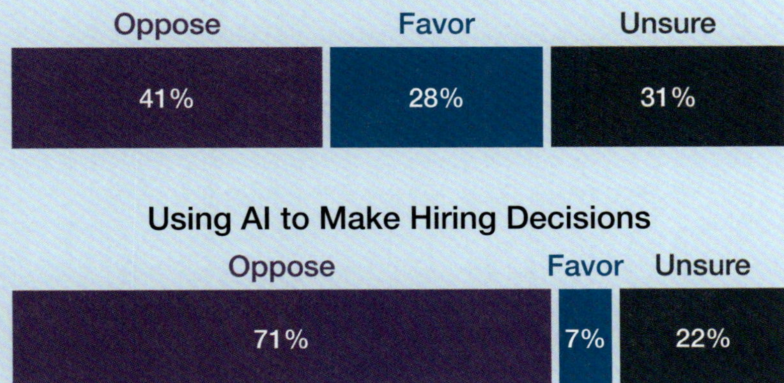


*Source: Lee Rainie, Monica Anderson, et. al. "AI in Hiring and Evaluating Workers: What Americans Think,"* Pew Research Center, *April 20, 2023. www.pewresearch.org.*

**Pew Research data indicates that while using AI to review résumés may be acceptable for some, the majority of Americans are against using the technology to decide who should be hired.**

to make predictions. These can be about weather or crime. AI can also predict supply and demand. This information can help people make faster decisions. For example, AI can help companies decide how much of a product to make.

# PERSONALIZING LIFE

Some people already have smart technology. This technology includes security systems, TVs, and appliances. In the future, these devices may be able to learn about their users. They could track people's behaviors and preferences. These devices could track people with

**Smart refrigerators of the future may provide recipe ideas based on the ingredients in the fridge.**

cameras, microphones, and sensors.
The data they collect could help predict a
person's wants and needs. For example,
AI can help people choose TV shows
or music.

AI has already turned some homes into
smart homes. These homes have smart
thermostats, doorbells, and more. Digital
assistants like Amazon's Alexa, Apple's Siri,
and Google's Bard already exist. They can
set timers, play music, and dim lights. AI
can help different devices communicate
with one another. This can make life more
**efficient**. For example, cameras on houses
could check for storm damage, pests, and
other issues.

Devices in smart homes could tailor
shows, products, and even meals

to individuals. This could save people time
and money. Michael Gardner is an architect.
He explains how smart technology is used
in his industry. He says, "It's such an integral
part of the home that we're designing it
from the beginning, where beforehand
technology was always an afterthought."[1]
This means smart technology will likely be
added to all homes in the future.

## IMPROVING FRIENDSHIPS

People have a need to connect with others.
Yet many people are lonely. AI could help
people make new friends.

AI already makes guesses about
relationships on social media. It suggests
friends. Social media platforms also show
posts from people with similar interests.

**On social media platforms, AI helps people make connections. It looks for data patterns to find other users who have similar interests.**

This connects people. But it can also be **polarizing**. This is because people spend most of their time seeing opinions that are like their own.

In the future, AI could use data to predict how well people will get along. In one study, AI focused on tone of voice. It correctly predicted relationship outcomes 79 percent of the time. Human therapists were right 76 percent of the time.

People can even use AI to create their own friends. They can design a chatbot's avatar. People can choose how it looks and sounds. AI can be programmed to think like its user. This may help the user feel more understood.

Chatbots will likely become even more personalized in the future. They might review messages, photos, and videos. This could help chatbots better understand their user. But this brings up privacy concerns. Some people may not want to share their personal information. People could be manipulated by the technology, too. Still, some predict that AI could become better at connecting with humans than other humans. Mike Brooks is a psychologist. He states, "People are falling in love with today's basic, entry-level chatbots, which have relatively limited capabilities compared to our future AI companions. When we do the math and connect the dots . . . we do not stand a ghost of a chance of resisting the allure and accessibility of these future AI companions."[2] AI may even replace friends and romantic partners for some people.

**AI makes it possible to create an avatar that has all the personality traits a user likes.**

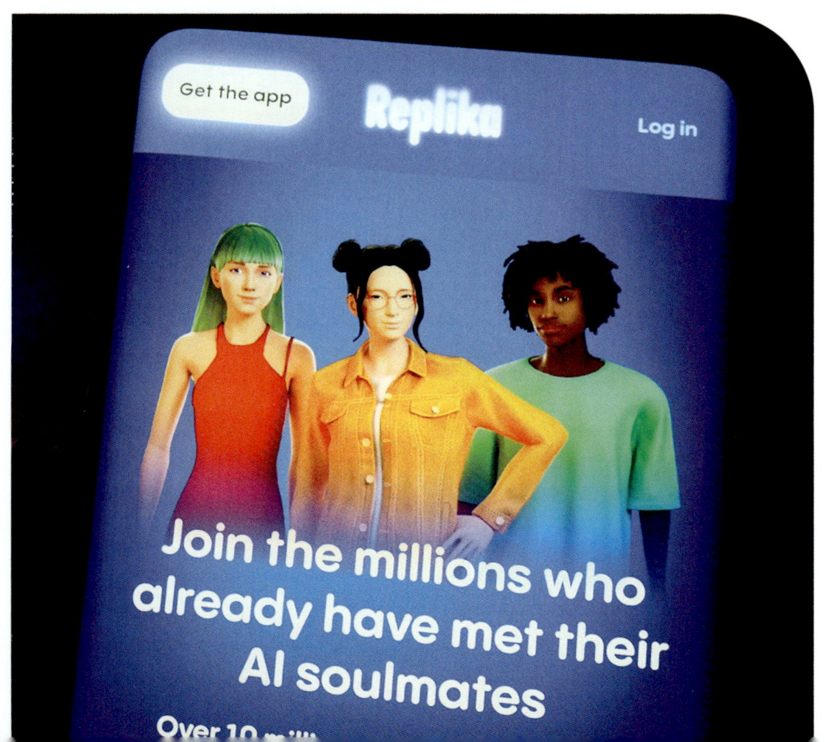

# CHAPTER TWO

# KEEPING PEOPLE SAFE

AI can keep people safer. People want to feel safe at home and while away. AI can give them this security. It can scan areas for threats.

AI already helps keep people safe. Some home security systems use AI. For example, the Ring security system includes a doorbell with a camera. It allows users to see people at their door. It tracks anyone who steps onto their property.

**People with home security systems can monitor their homes through an app.**

Ring also has an app for sharing security videos with neighbors. AI can be used to keep people safe in their communities and online. AI could make work and driving safer, too.

## COMMUNITY SAFETY

AI may be able to predict crimes before they happen. For example, AI uses data to predict which areas will have the

## Emergency Services

AI can also keep people safe during home emergencies. Smart home systems can detect threats such as smoke and carbon monoxide. They can alert homeowners of issues. They can also contact emergency services. The time saved by AI in these emergencies could save lives.

most crime. This can help police officers
decide where to patrol. It can help people
make decisions about whether to buy home
security systems.

AI could also be used to check for
weapons. It could use sensors and 3D
imaging to find them. AI could set off
an alert if a person has a gun where
firearms are banned. Jon Knight works
for a company that makes home security
systems. He says, "The benefit in being
able to detect openly held large firearms
would allow crucial seconds of lockdown,
emergency calling, or other forms of
protection that can save numerous
lives."[3] For example, AI could lock the
doors of a building when guns have been
detected outside.

# STAYING SAFE ONLINE

AI can keep people and information safer online. People are spending more time on the internet. This increases their risk of experiencing online attacks. AI can help protect people against cyberattacks. It can also protect them against false information and hate speech.

A cyberattack involves stealing or sharing private data. For example, someone might access people's bank accounts illegally. Cyberattacks rose 600 percent during the COVID-19 pandemic. This was partly because more people worked from home. Home computers tend to be less secure than those in offices. AI can be trained to detect threats. It can also help keep information safe. Smart technology networks in homes can be attacked, too. More homes will use this technology in the future. Smart devices will be exchanging

**As more people work, bank, shop, and complete other tasks online, strong cybersecurity systems play a key part in protecting personal data.**

more sensitive data. AI can help protect this data.

More people are getting their news from online sources. AI can be used to make sure this news is reliable. Many social media platforms already use AI. It checks for hate speech and fake news. It also looks for cyberbullying. Some video game companies plan to use AI to review voice chats. It will watch for harmful words, tone of voice, and increased volume. Reports will be submitted to a moderator.

## WORK SAFETY

In the future, AI may be used in dangerous jobs. This will help keep workers safe. AI-powered robots might be used to work with bombs and nuclear waste. AI could

**In the workplace, AI-driven robotic arms can be operated remotely, keeping workers safe from injury.**

also do jobs that can affect the health of workers. These include working with chemicals. AI can also do jobs that involve repetitive motions. This would reduce the risk of human injury.

AI can be more accurate than humans as well. The use of AI can reduce errors people make in their work. For example,

AI may detect errors in a process or product design. This could make work safer for workers. It would also make products safer for consumers.

## SAFETY ON THE ROAD

AI can keep people safe by reducing car accidents. Self-driving cars are driven by AI. Experts train AI to drive. They use computer software to **simulate** driving billions of miles. The software learns how it should react when driving. Self-driving cars use cameras, speakers, and sensors to help keep people safe.

Self-driving cars already use AI to plan their routes. States such as Arizona and California allow self-driving cars on their roads. Many experts think they will become a part of everyday life in the future.

Self-driving cars could keep people safer.

They would allow people to do other things

while traveling.

**Self-driving vehicles will make it possible for all passengers to read or watch a movie while safely reaching their destination.**

# STAYING SAFE DURING NATURAL DISASTERS

AI can find patterns in large sets of information. This can help it forecast natural disasters. AI can help predict earthquake aftershocks. It can share information about

**AI can be used to monitor increases in rainfall, wind, or drought conditions for a given area. This data helps people plan for the future.**

when and where these could occur. AI can also predict how strong earthquakes might be. In tests, AI has been able to do this more accurately than other systems. Phoebe DeVries has lead research in this area at Harvard University. She says, "We're still a long way from actually being able to forecast [aftershocks], but I think machine learning has huge potential here."[4]

AI can also predict floods. It can gauge current weather trends. It can track water levels in rivers and lakes. When a flood is likely, it can warn people to help them stay safe. This gives people more time to secure their homes and evacuate. It could save money and lives.

# CHAPTER THREE

# STAYING HEALTHY

AI is already used to keep people healthy. It will likely begin to play an even larger role in health care. Wearable devices could send patient data to doctors around the clock. This could make health care more effective. AI has shown promise in making many advances in the medical field.

AI can help doctors figure out what is wrong with a patient. It can help them

**AI can keep track of a person's heart rate, stress level, steps, and breathing during activity. This type of data can be used to help doctors monitor patients at home.**

find the best treatment. It can help
scientists create new medicines. It can
also help make tasks like record-keeping
more efficient.

## DIAGNOSING DISEASES

AI can be used to **diagnose** patients. AI
might even be able to predict the risk of
future health issues. It could do this by
using patient data. It could also use data
from patients with similar genetics and
health problems. AI may catch important
facts that health providers miss.

   AI is very good at finding certain
health issues. It can detect cancer earlier
than some other scanning options. It is
especially good at reading health scans.
These include scans that use magnets and

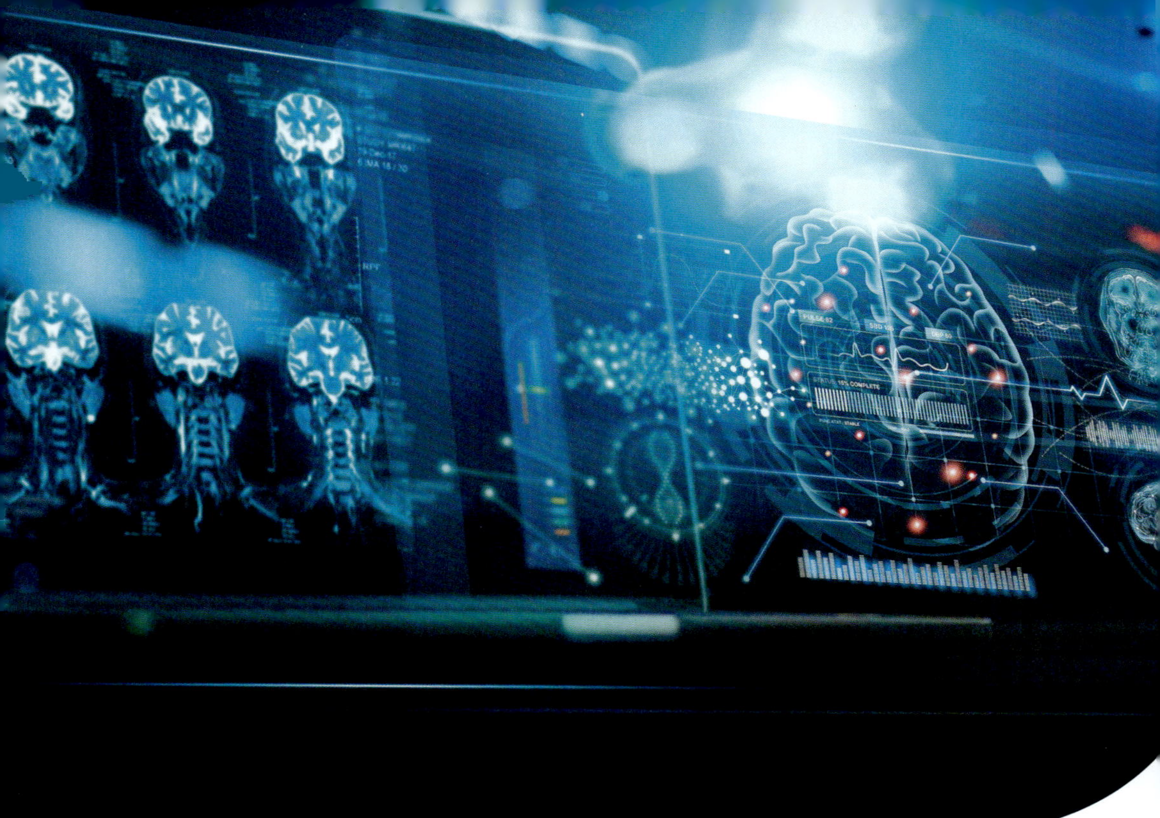

Doctors use AI programs to quickly study brain scans.
This can help them identify diseases earlier.

X-rays to take pictures inside a patient's
body. In some cases, AI is better than
doctors at reading scans. Isaac Kohane
is the head of Harvard Medical School's
Department of Biomedical Informatics. He
says, "I think [AI is] an unstoppable train in
a specific area of medicine . . . and that's
in image recognition."[5]

AI is trained using thousands of scans. Doctors study these scans. Then they label them with a disease. The AI program views a scan. It studies it. Then it identifies the illness and checks its answer. The AI program reviews more and more scans. It gets feedback on each one. AI improves with each scan it reviews. Still, AI can make errors. These errors could be difficult to detect.

Experts believe AI works best when it works with humans. For example, one study showed that AI could identify cancer. It was correct 92 percent of the time. Trained doctors found cancer 96 percent of the time. Then AI and doctors' opinions were combined. Cancer was correctly identified

99.5 percent of the time. AI could improve the accuracy of test results.

At home, smart mirrors could check for health problems. Mirrors could scan people's skin with cameras and sensors. Sensors in toilets could check waste for disease. They might discover cancers, diabetes, or infections. They could share this information with users and doctors.

## Chatbot Therapy

Doctors have created chatbots to help improve mental health. First, a chatbot tracks a person's emotions. Then, it asks questions to help the person feel better. Chatbots could be cheaper than human therapists. They could also be more available. Chatbot therapy combined with the care of a human therapist could be effective.

They could also help doctors track patients'
health. This could help patients avoid extra
trips to a clinic for lab work. Still, health
information is sensitive. Some people worry
about privacy.

## TREATING PATIENTS

AI could be used to help treat patients,
too. Researchers have studied this for
years. Yet much of what they've learned
has not yet been used in the field. AI
could study treatment options. It could
identify options that might work best. It
could even personalize treatment. AI could
do this based on a person's **genetic
makeup** and disease. Finale Doshi-Velez
is a professor of engineering and applied
sciences at Harvard University. She thinks

**Doctors use AI to find medical studies related to illnesses they are trying to treat.**

that if AI and doctors worked together, they could improve outcomes for patients. She explains, "I'm very excited about this team aspect and really thinking about the things that AI and machine-learning tools can provide an ultimate decision-maker . . . to empower them to make better decisions."[6]

AI could also help doctors perform surgeries. Surgical robots could help doctors make smaller incisions. They could also stitch patients' wounds. The robots could help doctors see inside patients. Providing doctors help with these tasks might make surgeries safer and more effective.

**Surgeons are learning how to work with AI-driven robotic arms in the operating room. These tools can improve precision.**

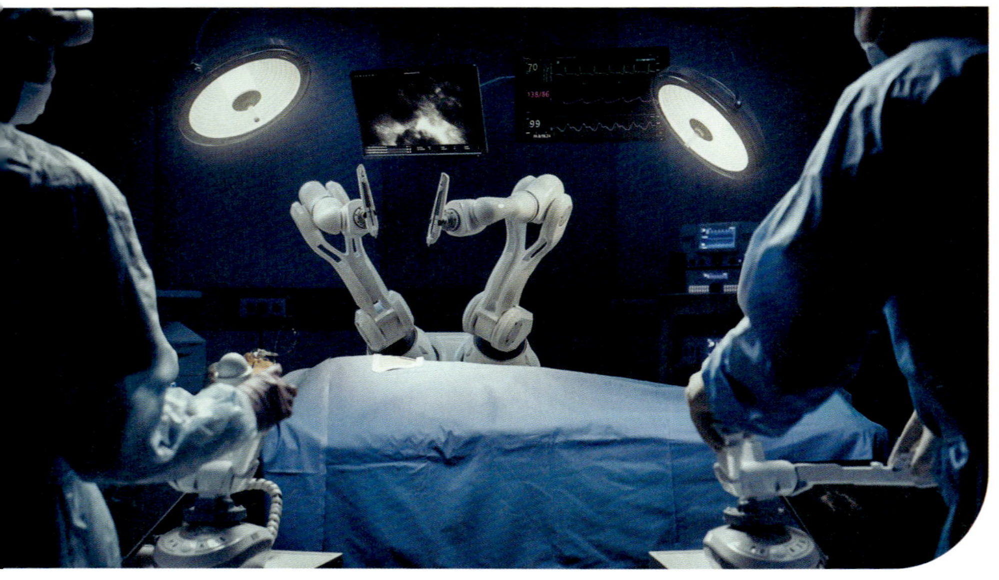

AI may be able to help experts create new medicines. AI can sort through medical studies. It can match a drug with a disease. It does this by aligning their molecules. Then AI can predict the results. This can help doctors quickly identify which drugs could work best for treating certain diseases. AI's ability to sort through data may also help it develop vaccines. It could predict how well different vaccines might work. This could prevent people from getting sick in the first place.

## PROVIDING BETTER CARE

AI could take on some office tasks in the medical field. This would give doctors more time to work with their patients. AI has the potential to reduce the amount

AI makes it more efficient for digital health records to be accessed and sent to new providers. This could improve patients' quality of care.

of paperwork that medical staff need to complete. It could also send health records to different hospitals when needed. This would give doctors access to patient information. AI could also schedule surgeries and help with billing. When work can be done faster and by fewer people, it could mean lower medical costs for patients.

Some experts believe AI could provide health care to more people. AI could also make it easier for people who live in rural areas to get care. It could even make care more affordable. But some worry that biases shown in AI could lead to inequalities in health care.

# CHAPTER FOUR

# PROTECTING THE ENVIRONMENT

AI helps protect the environment. The health of the planet has become a concern in recent decades. Taking care of the Earth can improve life for people and other living things.

AI could help humans reduce waste. It could also make waste removal more efficient. Sensors on garbage bins could track when they are full. They could alert waste management companies when bins

**AI could provide solutions for reducing waste and producing clean energy more efficiently.**

need to be emptied. This could reduce trips made by garbage trucks. This is just one way AI could lower energy use. AI could also help protect land, water, and animals.

## REDUCING WASTE

AI may help the world solve its waste problem. An average of one in four items placed in recycling bins are not recyclable. AI could help improve the recycling process. It can use computers to take pictures of waste. AI could use these images to sort trash. It could direct robotic arms to place waste into bins for paper, plastic, metal, and non-recyclable material.

Food waste is a large source of pollution. The United States alone wastes more than 100 billion pounds (45 billion kg) of food every year. This is equal to 30 to 40 percent of the country's food supply. AI **algorithms** can track how much food is purchased or used. This could help stores and restaurants predict how much food they need in the future. Gaining a better idea of the amount of food needed could reduce food waste.

**AI can help cut down on food waste by tracking how much food is needed. This can prevent food from ending up in dumpsters.**

# CREATING CLEANER ENERGY

AI could also help in the world's quest for cleaner energy. AI can use sensors to gather information about the weather.

**Wind turbine farms are one way for the world to transition to cleaner energy. AI can help determine where these farms would be most effective.**

It can use this data to predict weather patterns. People could use this data to make decisions about renewable energy. For example, it could help them decide where to build farms with wind turbines and solar panels.

AI can study how much energy a home has used in the past. It can compare this information with weather data. AI can then predict how much energy will be needed to power the home in the future. This can help plan energy needs. A home could store extra energy in a battery. This could be used later instead of being wasted.

## CONSERVATION EFFORTS

AI can also help with conservation. It can track water loss through satellite images.

For example, Brazil lost over 15 percent of its surface water between 1985 and 2020. This affects the land and the animals that live there. Cássio Bernardino is a project lead with World Wildlife Fund. He says, "AI technology provided us with a shockingly clear picture. Without AI . . . we would never have known how serious the situation was, let alone had the data to convince people. Now we can take steps to tackle the challenges this loss of surface water poses to Brazil's incredible biodiversity and communities."[7]

AI can also help protect animals. AI helps park rangers in Africa watch for poachers. The AI has been trained on what to look for. It uses heat sensing cameras to track people and vehicles in restricted areas.

Satellite images taken each year show scientists how
the environment is changing.

**Using heat-sensing cameras, AI can capture poachers on video when they enter off-limits areas. This could help park rangers protect endangered animals.**

It alerts rangers to threats. It also helps them watch over large areas of land at one time.

AI also helps track animal species. It uses cameras, satellite images, and audio recordings. Tracking animals helps humans learn how to better support them.

AI has the potential to help people live better lives. It can save people time and keep them safe. AI can improve the health of people and the world.

## Better Understanding Animals

Some scientists think AI will help people better understand animals. AI can help scientists decode animal calls and behavior. Christian Rutz is a behavioral ecologist. He says, "We are on the brink of fairly major advances in regard to understanding animals' communicative behavior."

*Source: Quoted in Lois Parshley, "Artificial Intelligence Could Finally Let Us Talk with Animals,"* Scientific American, *October 1, 2023. www.scientificamerican.com.*

# GLOSSARY

**algorithms**

sets of rules used to solve problems

**diagnose**

to identify an illness or health issue

**efficient**

done in a way that maximizes effectiveness and minimizes waste

**genetic makeup**

characteristics passed down through family genes

**machine learning (ML)**

AI that allows systems to learn and improve from experience

**polarizing**

causing strong disagreement between different groups of people

**sensors**

devices that take in data about their surroundings

**simulate**

to make up a situation that allows people to practice a skill they will use in real life

# SOURCE NOTES

### CHAPTER ONE: MAKING LIFE EASIER

1. Quoted in Patrick Lucas Austin, "What Will Smart Homes Look Like 10 Years from Now?" *Time*, July 25, 2019. www.time.com.

2. Quoted in Mike Brooks, "Future Love: How AI Companions Will Capture Our Hearts," *Psychology Today*, April 11, 2023. www.psychologytoday.com.

### CHAPTER TWO: KEEPING PEOPLE SAFE

3. Quoted in Ashley Brooks, "The Benefits of AI: 6 Societal Advantages of Automation." Rasmussen University, November 4, 2019. www.rasmussen.edu.

4. Quoted in James Vincent, "Google and Harvard Team Up to Use Deep Learning to Predict Earthquake Aftershocks," *Verge*, August 30, 2018. www.theverge.com.

### CHAPTER THREE: STAYING HEALTHY

5. Quoted in Alvin Powell, "AI Revolution in Medicine," *Harvard Gazette*, November 11, 2020. www.news.harvard.edu.

6. Quoted in Powell, "AI Revolution in Medicine."

### CHAPTER FOUR: PROTECTING THE ENVIRONMENT

7. Quoted in Graeme Green, "Five Ways AI is Saving Wildlife—from Counting Chimps to Locating Whales." *Guardian*, February 21, 2022. www.theguardian.com.

## BOOKS

Stuart A. Kallen, *Changing Lives Through Artificial Intelligence*. San Diego, CA: ReferencePoint Press, 2021.

Jennifer Kaul, *The Dangers of Artificial Intelligence*. San Diego, CA: BrightPoint Press, 2025.

Betsy Rathburn, *Artificial Intelligence*. Minneapolis, MN: Bellwether, 2021.

## INTERNET SOURCES

"Artificial Intelligence," *Britannica*, n.d. www.kids.britannica.com.

Nik Popli, "This AI Job Is Hot," *Time for Kids*, May 2, 2023. www.timeforkids.com.

Stephanie Warren Drimmer, "Could a Robot Become President?" *National Geographic Kids*, n.d. www.kids.nationalgeographic.com.

# WEBSITES

### Google AI
**www.ai.google**

Google is a large technology company. Its website shares information about its AI products.

### OpenAI
**www.openai.com**

OpenAI is the company that developed ChatGPT. Its website explains the uses for OpenAI products and provides information about types of AI. The site also includes articles about OpenAI research.

### PBS Crash Course: Artificial Intelligence
**www.pbs.org/show/crash-course-artificial-intelligence/**

Crash Course: Artificial Intelligence is a series hosted by Jabril Ashe that teaches viewers about AI.

# INDEX

# IMAGE CREDITS

# ABOUT THE AUTHOR

Jennifer Kaul is an author of children's and young adult literature, a former educator, and a cautious optimist. Many of Kaul's pieces stem from the happenings in our world and the *what-ifs* that swirl around her head as a result. Through her writing, she hopes to encourage thought, spark conversation, and make the world a better place.